## A PRACTICAL GUIDE FOR MANAGING THE ASIAN HORNET

**A Practical Guide For Managing The Asian Hornet**


Published 2024 by
Northern Bee Books,
Scout Bottom Farm,
Mytholmroyd,
West Yorkshire
HX7 5JS (UK)
Tel: 01422 882751
Fax: 01422 886157
www.northernbeebooks.co.uk

ISBN 978-1-914934-74-2

Design and artwork DM Design and Print

# A PRACTICAL GUIDE FOR MANAGING THE ASIAN HORNET

## Contents

# 1
# Introduction

Asian Hornets are here to stay, but it's not the end of the world.

It's not the end of beekeeping.

But it could be the end of beekeeping as we presently know it.

For our bees to survive the ravages of predation by Asian Hornets, and to continue to thrive, we may need to change the way we keep our bees and alter our expectations of what they can do.

This means taking command in our apiaries, reducing the level and stress of predation, and ensuring that the colony continues to function as normally as possible whilst predation is in progress.

My first experience of Asian Hornets wreaking havoc was in my modest apiary in the Loire Valley of France. I learned a great deal from my friend Monsieur André Blatier, a shrewd, weather-beaten old *Maître apiculteur.* He taught me that, by putting in place a system of Integrated Apiary Management, we can survive the Asian Hornet problem, get our bees through the predation period and into winter, and still have a crop of honey in return for our efforts.

To achieve this, we have to be very close to our colonies and sharpen up our beekeeping skills and apiary procedures. To put it simply, we need to raise our game and start thinking out of the box.

This book proposes a set of simple principles with guidelines for we amateur beekeepers to help manage our bees and apiaries when they're under attack. We aim to ensure that our investment of time, effort and money is still worthwhile in the new situation.

*Your very first steps are, if you haven't already done so, to complete the BBKA Asian Hornet Recognition Exercise and to download the Asian Hornet Watch App on your mobile phone.*

To read more about these dangerous but fascinating insects, try Devon beekeeper Dr Sarah Bunker's excellent book *'The Yellow-Legged Asian Hornet - Biology, Spread and Control'. (Published 2024)*

Learning about the life cycle of the Asian Hornet is the starting point for understanding and planning our response to this new threat to the health and wellbeing of our colonies of honey bees.

# 2
# Know the Enemy

*"Shape your plans according to the rule and circumstances of your enemy."*
*The Sayings of Wu Tzu*

To understand how we can successfully react to the threat posed by Asian Hornets to our precious bees, it's essential to consider how they live.

This simple diagram shows the key stages in the life of the Asian Hornet. The traffic-light colours correspond to the level of threat to honey bee colonies and to our responses:

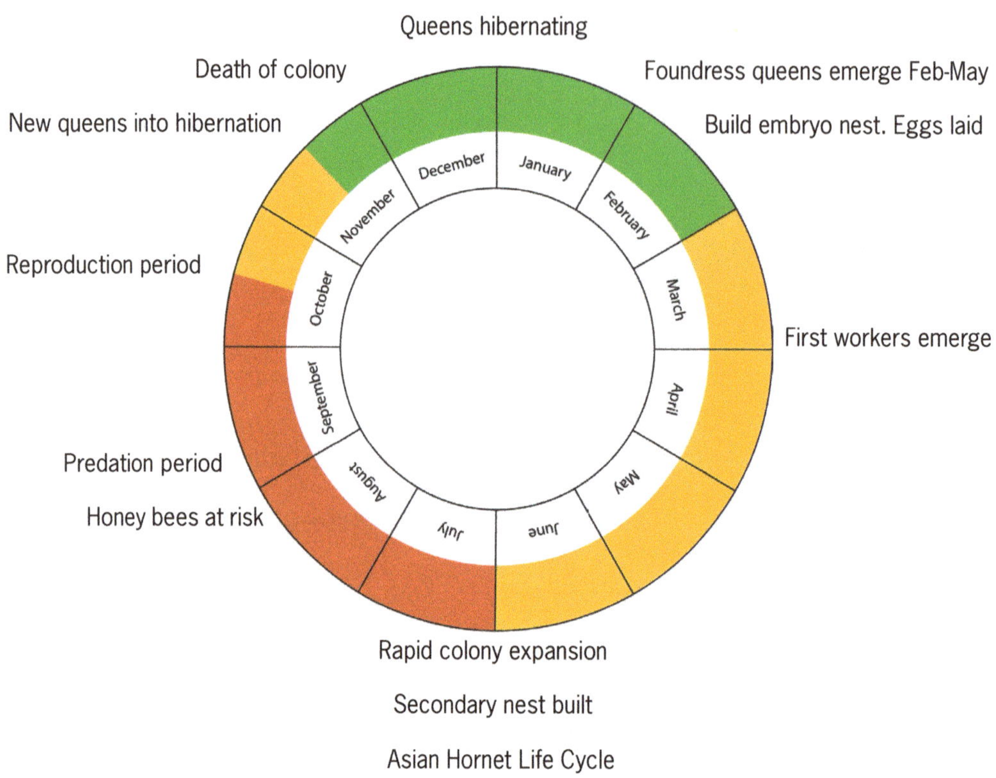


Asian Hornet Life Cycle

## 2.1 A Phased Response

*"Victory can be foretold when the general knows the time to fight and when not to fight."*
*Sun Tzu: The Art of War*

We can give the bees a better chance of riding the storm by adjusting the timing of our beekeeping calendar to complete as many tasks in the colonies as possible before predation begins, then progressively locking down the hives until the danger has passed. The same traffic light coding as in the AH life cycle can be used in our response:

**Green** I'm in an area where no hornet nests have been found yet.

**Amber** Asian Hornets confirmed in my area.

**Red** Asian Hornets have been seen actively hunting in or near my apiary.

## 2.2 Effects of Predation

Asian Hornets hunt on the wing by hovering ("hawking") in front of the hives, usually with their back to the entrance. They have a wide view of returning foragers, which are easier prey as they fly slower when laden with nectar, pollen, or water.

Colony under attack, bees clustered in front of the hive entrance.

*Image courtesy of Michael Judd*

The unlucky victim is carried to a nearby branch for dismembering. The big protein-rich flight muscles in the bee's thorax are removed and taken back to the hornet's' nest where they're fed to the begging larvae. In return, the larvae exude a rich 'superfood' which encourages the foraging worker to go back for more prey.

Asian Hornet dismembering a bee

*Image courtesy of Michael Judd*

When a colony is under sustained attack the bees refuse to leave the hive, preferring to cluster at the entrance in a defensive wall. If the colony is strong, the hornets hesitate to land near the entrance and avoid confrontation with the colony's guard bees. So far, so good

However, the terrified bees stop foraging for nectar, pollen and water in what is known as foraging paralysis. They become stressed, exhausted, and less able to defend the hive. As normal activities cease, the queen is likely to stop laying which, together with the lack of foraging, means the colony will have insufficient winter bees and stores to survive until the following spring.

The colony becomes progressively weaker, there are no guard bees left to defend the entrance, and the emboldened hornets are free to enter the hive and plunder what's left. Unlike wasps, the Asian Hornet takes everything: bees, larvae, honey and even comb!

The main predation period is in mid to late summer and autumn. The actual timing depends on geographical location, climate, and the weather at the time. It usually begins in July and has been known to extend into November in some areas. This is the time when we need to be at maximum vigilance.

We can support our bees by adjusting the timing of the beekeeping calendar, completing as many tasks as possible in the colonies before predation begins and creates stress with all its consequences. The colonies can then be put into 'lockdown' until the danger has passed.

*NB If a queen is damaged or lost during an inspection once predation has started, there's little chance of the colony being able to replace her.*

# 3

# Preparing the Troops for Battle

## Three Pillars of Fit2Fight

When the future existence of any organism - animal, plant or insect - is under threat it has a better chance of survival if it is fit, strong and well-nourished. With better understanding of the timing of the predation, we can take steps to ensure that the three pillars of the Fit2Fight principle are in place when they will be most effective for our frontline bees. The 3 pillars are:

**Healthy bees**

**Strong colonies**

**Well-fed stock**

# 4

# F2F Healthy Bees

*"A healthy soldier is a ready soldier."*

*Lt Gen Scott Dingle*

Every time you inspect a colony, ask these 5 questions:

- Is the queen present, either seen or signs of presence through all stages of brood?
- Is the colony building up since the last inspection and compared with other colonies?
- Are there any signs of disease ?
- Does the colony have sufficient space ?
- Does the colony have sufficient stores ?

In addition, the following four elements can help to assure good health in the apiary:

- Health inspections
- Low Varroa load
- Strict apiary hygiene
- Regular comb change.

## 4.1 Health Inspections

Detailed inspections for the presence of diseases or parasites affecting adult bees and brood begin in spring soon after the first inspections of the season and continue monthly until the beginning of predation.

A good health inspection requires a separate visit to the colony in its own right and shouldn't be combined with any other task.

Knowing what to look for is essential as the signs and symptoms of some diseases are not always obvious. Attending Bee Health Days and doing the BBKA Honey Bee Health Certificate are good ways of learning and improving knowledge of honey bee pests and diseases.

Take photos and keep a record of each inspection.

Watch the bees at the hive entrance, and inside the hive, and look out for any abnormal appearance or behaviour.

During each health inspection we're looking for signs of unhealthy bees and brood such as:

- Malformed workers.
- Deformed wing virus.
- Scattered brood pattern.
- Perforated sealed brood.
- Sunken cappings.
- Greasy looking cappings.
- Scales on the bottom of cells
- Mal formed larvae.
- Discoloured larvae (not pale creamy-white).

Shake the bees off every frame one by one and examine the brood carefully, looking out for any abnormality.

Examining a frame of brood

*Courtesy the Animal and Plant Health Agency (APHA), Crown Copyright*

A frame of healthy brood

*Courtesy the Animal and Plant Health Agency (APHA), Crown Copyright*

Deformed Wing Virus

*Courtesy the Animal and Plant Health Agency (APHA), Crown Copyright*

Chronic Bee Paralysis Virus

*Courtesy the Animal and Plant Health Agency (APHA), Crown Copyright*

If anything unusual is seen or suspected, begin remedial measures straight away or *seek advice if you're not sure what to do. Don't try and manage it yourself.*

If contagious diseases such as CBPV or DWV are found, put the hive in 'quarantine' and follow the treatment guidelines as described in the BBKA Healthy Hive Guide* until the symptoms have disappeared, and the colony has recovered.

If you suspect you might have American or European Foulbrood you must immediately:

- Stop the inspection.
- Close the hive entrance to one bee space.
- Call the Bee Inspector.
- Put the apiary in voluntary stand fast until the Inspector arrives.
- Destroy your gloves.
- Sterilize your hive tool and other equipment.
- Wash your bee suit and all your clothes in soda crystals and soap solution.

American Foul Brood (AFB)

*Courtesy the Animal and Plant Health Agency (APHA), Crown Copyright*

European Foul Brood

*Courtesy the Animal and Plant Health Agency (APHA), Crown Copyright*

*The BBKA *Healthy Hive Guide* is a handy pocket reference for adult bee and brood diseases.

For more on the subject see *Healthy Bees are Happy Bees by Pam Gregory*

## 4.2 Low Varroa Load

*Varroa destructor* is aptly named. This horrible little beastie is a major cause of failure in our colonies and is associated with a wide variety of brood and adult bee diseases, including Deformed Wing Virus, Chronic Bee Paralysis Virus and more, leading to absconding or colony collapse.

Varroa on a larva

*Courtesy the Animal and Plant Health Agency (APHA), Crown Copyright*

Varroa mites prefer to lay their eggs and rear their young in drone brood. The longer development of drones (24 days as opposed to 21 days for workers) gives the varroa time for an extra brood cycle allowing the female to lay more eggs and produce more adults.

The varroa mites feed on the haemolymph of the bee larvae. In the case of the drone, this has consequences for the adults, including the quality of their semen. Poor sperm can cause reduced fertility in queens, weaker offspring and early queen supersedure.

- By practicing a rigorous programme of varroa management, we can keep the level of infestation as low as possible. The management steps include the use of chemical and non- chemical (sometimes known as biotechnical) methods:

- Regular varroa load testing by counting the drop on the varroa boards and by sugar roll or alcohol wash. A rate of a 1,000 mites per colony is the maximum.
- Drone brood uncapping which removes a lot of the varroa eggs, larvae, pupae and young adults.
- Comb trapping is an effective technique for *Varroa* control; however, some skill and extra equipment are needed in carrying out the technique. It is claimed to give an efficacy of 95% and one of the advantages is that no chemicals are used when supers are on the hive. For more information visit https://shorturl.at/HSU09

Varoosis

*Courtesy the Animal and Plant Health Agency (APHA), Crown Copyright*

Honey bee with varroa mite on its abdomen

- Comb change by shook swarm. Some brood is lost but the colony is given a varroa -free fresh start to the season and colonies enjoy a huge boost in population afterwards.

Shook swarm

*Courtesy the Animal and Plant Health Agency (APHA), Crown Copyright*

- Oxalic acid treatment in winter, just before the solstice (about the 21st December), when there is little or no brood, and in spring after shook swarm. This can be carried out by trickle, GasVap or sublimation (often called vaporizing or fumigating) in which solid oxalic acid is transferred directly into a gaseous state.

- Treatment with miticides such as Formic Pro (formerly known as MAQS) which only requires one dose applied after the honey harvest has been taken and thus reduces the number of times the hives are opened. *
- Collected swarms which have been placed in isolation are given Oxalic Acid treatment while they're still without capped brood.

Beekeepers attempting to develop treatment-free colonies may find themselves vulnerable to attack by Asian Hornets in the early years and experience heavier losses than they would normally expect.

* The occasional problems with loss of queens reported by some beekeepers using MAQS can generally be avoided by having strong, healthy colonies with vigorous young queens, and by following the manufacturer's instructions to the letter. I have been using MAQS for several consecutive years without any problems or signs of resistance. Just in case, I do a Beltsville Resistance Test on two or three sample colonies every year to check.

## 4.3 Strict Apiary Hygiene

It is essential to prevent the introduction and spread of infections and parasites.

Many diseases are carried and transmitted by beekeepers. A vital part of keeping our bees healthy is maintaining the apiary and all our equipment clean and free from contamination.

Don't allow used comb, eggs, larvae or bees from other apiaries to be used or introduced into your colonies.

If you wear gloves for inspections, reduce the transmission of nasties between colonies by using disposable ones and changing them between each hive. If you wear Marigolds rinse them in soda crystals and soap solution between hives.

Leather gloves are clumsy, can aggravate the bees, and are a rich breeding ground for infection so are best avoided. If you need to wear leather gloves for protection, you probably need to requeen the colony to have calmer, less defensive bees.

After apiary visits wash bee suit and other clothing in soda crystals and soap.

Hive tools should be scrubbed with wire wool to remove wax and propolis, then rinsed in soda crystals and bleach solution between each hive inspected.

The bellows of smokers can be protected and kept clean with a disposable shower cap.

Hive debris such as wax and propolis scrapings should be put in a lidded container and remove from site.

The ground should be kept clear of any debris, especially edible matter such as fallen fruit.

Used wooden equipment should be scraped free of wax and propolis and sterilized with a blow torch before being stored or reused. Poly equipment can be scrubbed with wire wool and immersed in a mixture of soda crystals, bleach, soap and hot water.

Super or brood frames should be put in the freezer at least -18°C, or fumigated with 80% Acetic acid to kill wax moth and other parasite eggs, pupae, and other pathogens that are lurking in the comb.

## 4.4 Regular Comb Change

Dirty brood comb can contain an accumulation of old larval skins, residual chemicals from treatments given by the beekeeper and brought in by foragers, and more. It's the perfect breeding ground for all sorts of nasty gremlins. In normal times we are encouraged to change brood comb at least every 3 years.

The arrival of the Asian Hornet increases the importance of keeping healthy bees, and more frequent comb change is a component in maintaining disease-free stocks.

Annual or two-yearly comb change by shook swarm, followed by treatment with Oxalic acid, can be a multiple winner by:

- removing accumulated pathogens and keeping the bees in a clean, healthy environment.
- boosting population growth
- preventing swarming
- reducing varroa load

## 4.5 Collected Swarms

If you collect, or are given, a swarm it should be kept in isolation until after a complete brood cycle in case it's infected with disease or parasites.

At the end of quarantine, a health inspection should be carried out to check for disease, followed by treatment with Oxalic acid before any brood is capped to reduce the varroa level.

# 5
# F2F Strong Colonies

*"Superiority of numbers is the most common element in victory".*
*On War: Carl von Clausewitz*

Like most life forms, numerically strong colonies are better able to withstand attacks by pests and diseases, than weak ones. Before the arrival of Asian Hornets, most of us were used to the population of our colonies increasing gradually from spring onwards to achieve a maximum foraging force in July in time for the main nectar flow.

Unfortunately, that main summer flow may coincide with the start of the Asian Hornet predation period. If the bees can't fly in summer, they must have achieved maximum income of nectar and pollen before predation starts, and we will have to supplement their food and water stores until the threat has retreated.

Therefore, we need rapid population growth in spring to take advantage of the spring early summer nectar flows. The problem is that this comes at a time when there is an increasing amount of brood and the winter bees are coming to the end of their lives. This is why the previous summer and autumn feeding is important to keep the queen producing strong winter bees with a long life expectancy for the following Spring.

Those of us who take colonies to fields of winter-sown Oil Seed Rape will be familiar with these techniques:

## 5.1 Spring Colony Building

**Aim:** To build the colonies to field a maximum possible foraging force in April and May

**Requirements**

- Productive young queens marked (and clipped).
- Healthy colonies with low varroa loads.
- Prevention of swarming – high risk during rapid spring build-up.
- Equal-sized colonies for simplicity of management
- Large quantities of bees emerging at the end of February/early March to become foragers in April and May
- Brood nest rearranged with sealed brood at the centre and unsealed on the outside to discourage bees from putting honey in the brood box.
- Supers ideally with frames of drawn comb added once foraging begins.

## Potential risks

- Swarming
- High ratio of brood to adult bees as the winter bees come to the end of their lives.
- Workers eating eggs to limit growth of colony if no flow or insufficient bees to feed larvae (worker policing).
- Presence of Nosema limiting spring growth

| Timing | Action | Notes |
|---|---|---|
| **Autumn** | Ensure colonies are healthy, strong and well fed going into winter. Feed pollen if in short supply. Feed syrup to top up stores.<br>Introduce a new, young queen. | Varroa treatment applied after honey taken off to protect winter bees |
| **Winter** | Ensure colonies have stores by hefting | Oxalic acid treatment when little or no brood |
| **End Jan** | Start continuous feeding sugar syrup, and pollen if needed, to simulate early nectar flow and stimulate queens to lay.<br>If weather is bad, they may not come up and over a feeder so may need to feed fondant instead. | NB There is a danger that if the feed syrup is in the brood box they could move it up into the supers to give the queen room to lay. So, ensure plenty of room for her to lay. |
| **End Feb** | Spring inspection when temperatures permit<br>Continue to feed sugar syrup (and pollen if needed). | Check stores, space for queen to lay, change floor, check varroa drop. Treat for varroa if >1,000 per colony |
| **Early Mar** | First colony inspection<br>Health inspection<br>Check queen present & laying.<br>Inspect weekly for signs of swarm preparation.<br>Continue to feed sugar syrup | Check queen present & laying. Change queen if not performing.<br>Take samples to check for Nosema if build up slow.<br>Swarm prevention measures if required |
| **Mid Mar** | Weekly inspections<br>Equalize colonies *see below<br>Inspect weekly for signs of swarm preparation<br>Continue to feed sugar syrup | Monitor colony growth<br><br>Swarm prevention measures that will not interfere with the honey production if required |
| **End Mar** | Inspect weekly for swarm prevention<br>Continue to feed sugar syrup | In theory, if queens are clipped, inspections can be every 10 days, but the last thing you want is for them to swarm, return, but you lose the queen. It's better to carry on with weekly inspections just in case. |
| **Early Apr** | Ensure queen has room to lay<br>Continue to feed sugar syrup<br>Add supers<br><br>Weekly inspections for swarm prevention essential.<br>Swarm control carried out immediately the first charged queen cups are found. | Brood boxes can rapidly run out of space as the nectar flow is abundant and the colony might swarm. Move full frames to the outside, put empty frames near the centre. Adding an extra brood box gives more space and the bonus of getting some drawn comb and it gives the growing population of bees something to do |

* Strong colonies can be held back to reduce congestion, with consequent risk of swarming, and weak ones strengthened by transferring brood between them.

However, giving brood alone to a weak colony may not work if there are too few young nurse bees to look after it, so proceed as follows:

- Place a super of drawn comb or foundation on the strong colony, without a queen excluder, and allow the queen to lay in it.
- When the super is full of brood and bees, smoke the bees and the queen down into the brood box, ensuring that there's enough room for the queen to continue to lay.
- Put on the queen excluder and check after a few days that there are no eggs in the super and therefore the queen is in the brood box.
- Put the super on the weak colony which will soon have a boost in population of young bees.

## 5.2 Well-insulated hives

If hives are well-insulated, the colony will have used fewer resources and less energy keeping warm through winter, the internal temperature of the brood nest will be higher and the queen may start laying earlier.

The winter bees will be more rested, have more resources left in their fat bodies and therefore be better able to look after spring brood when the brood population exceeds that of the adult bees.

See recent work by Derek Mitchell of Leeds University on the benefits of all-year-round insulation.

## 5.3 Early Spring Feeding

Feeding syrup and pollen supplement or pollen substitute in very early spring stimulates the queen to start or increase laying.

In the case of honey-producing colonies, feeding should stop as soon as the spring flow starts to avoid contamination of the honey with sugar syrup.

See 6 below

## 5.4 Rigorous swarm prevention and control

As part of the Fit2Fight programme we're asking our queens to produce a lot of new bees in the early part of the year.

Swarming is a serious risk during rapid spring build-up, so it is vital to understand and recognize the early signs of swarm preparation and to watch them very carefully.

With the threat of Asian Hornets hawking in our apiaries in a few short weeks' time, we can't afford to lose any bees. If we do, there is less chance of building numbers up for the spring flow and going into the predation period with a big, strong colony.

In order to deal with the ever-present threat of losing half your bees and terrorizing your neighbours, it's useful to understand the swarming process, the key indicators to look out for and the prevention or control measures you can take.

In the flow diagram below you will see the progression that occurs. I've divided it into two parts: the times when you can take **prevention** measures ; times when **control** is necessary.

### Swarm Prevention

Prevention requires regular, thorough inspections and careful observation, at least weekly.

Here are some simple but effective measures you can take:

### Mark your queens

- If you can't find the queen get another pair of eyes or two to help you.
- The earlier in the season the better when there aren't too many bees.
- Do it in the warmest part of the day when the maximum number of bees are out foraging.
- Or move the brood box to one side and put a super on its original stand to divert all the flying bees and give you more room and quiet to look.

**Clip your queens.**

- This involves gently cutting off the end third of one wing.
- It isn't to everyone's taste but there are no nerves or blood vessels there – it's like cutting your toenails.
- If you're struggling with any of these, get a more experienced beekeeper to help you. People are always happy to lend a hand.

**Add space**

- The bees need room to live and to store incoming nectar and pollen.
- Nectar contains about 80% water and requires a greater volume of space than honey which only contains less than 20%.
- The queen needs more space to lay her eggs.
- Add a super as soon as the brood box starts to look crowded or they start to build comb on top of the brood frames. When the flow starts they will be bringing in nectar very fast.
- If there are no supers they will store nectar in the brood box, thereby depriving the queen of laying space and she will stop laying.
- Remember that during the day, a lot of the bees will be out foraging. At night they need somewhere to sleep.
- Replace surplus frames of honey with drawn comb or foundation.
- Replace damaged comb which can't be used efficiently.
- Be prepared by having spare equipment ready in advance to take off the pressure when the time comes to act.

## Swarm Control

Once queen cells have been formed, the time for prevention has passed and control measures are needed.

I prefer to use either the Nuc or the Pagden method if I want to make increase from the colony, or the Demaree technique if I don't want any more hives or to breed from that particular queen.

All three can be found on my Blog https://www.alanbaxtersblogs.co.uk/, numerous books, websites and on YouTube.

## Beware the cast, or secondary swarm

After the first swarm has issued, the sealed brood that the queen had been producing beforehand will start to emerge (remember those big slabs of brood about 3 weeks ago). To reduce the risk of secondary swarming:

- Remove all the sealed queen cells.
- Choose the best two unsealed ones, marking their position on the frame with a drawing pin.
- One week later go back and remove any more queen cells that have been made and
- remove one of the marked ones
- Make notes in your hive records of the dates – it can take longer than you think for a new queen to be mated and start to lay.

Don't be beguiled into thinking that your colony hasn't swarmed because the hive is still full of bees. It could be all that sealed brood which has emerged after the main swarm has left!

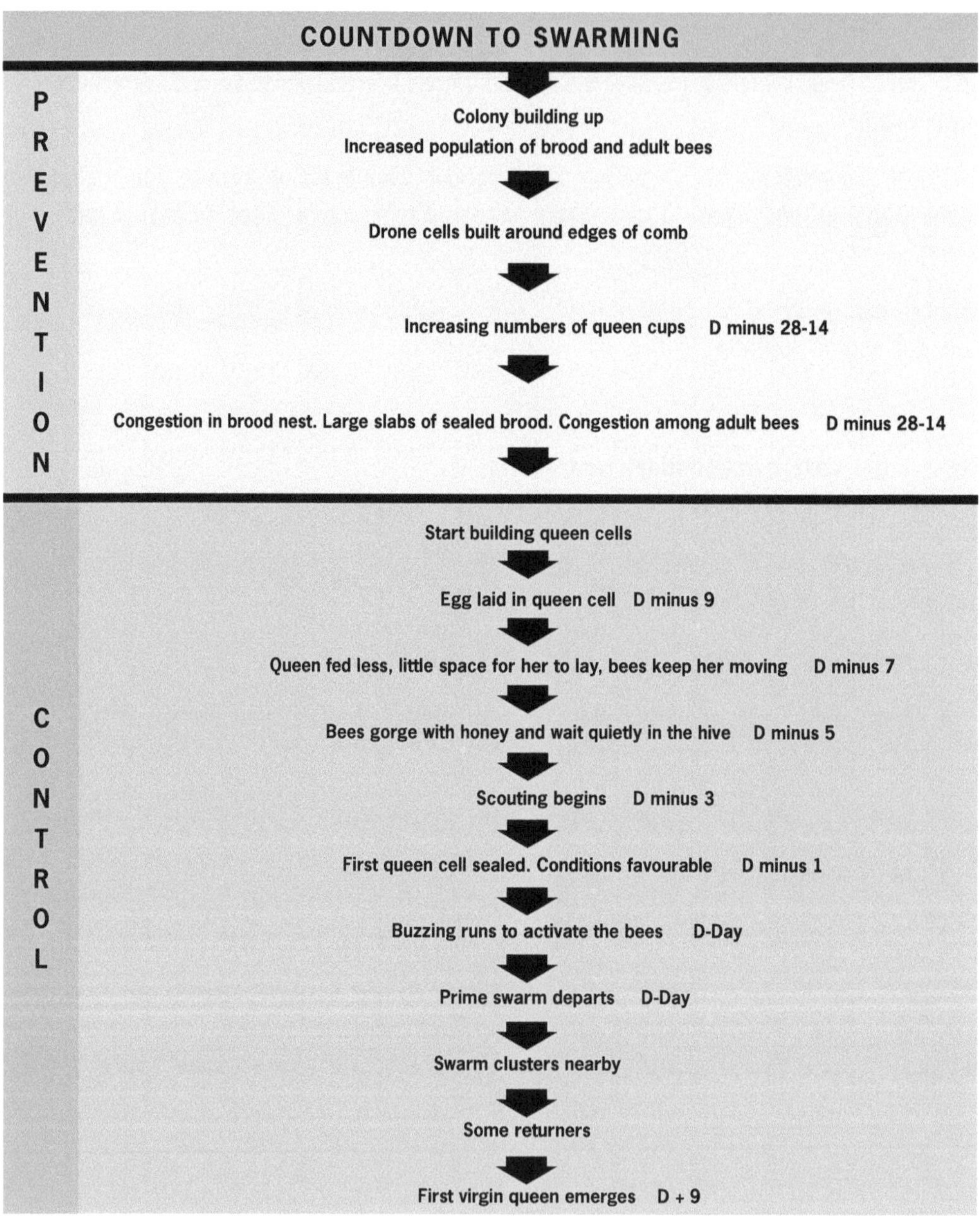
COUNTDOWN TO SWARMING
PREVENTION
Colony building up
Increased population of brood and adult bees
Drone cells built around edges of comb
Increasing numbers of queen cups D minus 28-14
Congestion in brood nest. Large slabs of sealed brood. Congestion among adult bees D minus 28-14
CONTROL
Start building queen cells
Egg laid in queen cell D minus 9
Queen fed less, little space for her to lay, bees keep her moving D minus 7
Bees gorge with honey and wait quietly in the hive D minus 5
Scouting begins D minus 3
First queen cell sealed. Conditions favourable D minus 1
Buzzing runs to activate the bees D-Day
Prime swarm departs D-Day
Swarm clusters nearby
Some returners
First virgin queen emerges D + 9

## 5.5 Uniting weak colonies

One strong colony has more than double the chance of surviving predation than two weak ones.

Uniting colonies in spring enables them to field a bigger foraging force for the spring flow and go into predation lockdown with lots of bees armed with plenty of stores.

Colonies that are struggling to increase numbers during May and by mid-June should be united in plenty of time before predation starts. Bear in mind that complete uniting takes a full brood cycle of 24 days.

It is worth asking yourself why they are weak at this time of the year. For example:

- Was the build-up normal in the spring ?
- Do they have a disease such as Nosema which is preventing them from growing? If in doubt arrange for a sample of bees to be tested.
- Could they have swarmed and you didn't notice?
- Has the queen been superseded ? If all your queens are marked you will know.
- Is it a problem with the queen? Does she need to be replaced?
- When did you last do a full health inspection? Was everything normal?
- What is the varroa count? Do you need to treat them now rather than wait until July?

## 5.6 Young productive queens

In order to produce the colonies best suited to the task of fighting the Asian Hornet, ideally queens will have been selected for:

- Rapid spring buildup
- Low swarming characteristics
- Hygienic behaviour for disease resistance.

Whilst this might not be feasible for many hobbyists with small apiaries and limited time. Local BKAs, or groups of individuals, can help by supplying suitable queens.

However, if you need to buy in queens, those raised in other areas will adapt less well to your local conditions. It's much better to find one from a local breeder or from your Beekeepers' Association.

Marked queen

*Courtesy the Animal and Plant Health Agency (APHA), Crown Copyright*

## 5.7 Queen rearing timing

Queen rearing needs to be concentrated into spring and early summer and finished before predation starts. Giving colonies drone foundation in spring has a triple benefit of:

- Stimulating queens to produce drones for early queen rearing.
- Improving the population of the local drone congregation area with "good" drones.
- Allowing uncapping of drone brood to remove varroa as part of Integrated Pest Management once queen rearing is finished.

Mating nucs are extremely vulnerable to predation so new queens should be introduced to full colonies, or 14 x 12 nucs with 6 frames, as soon as possible after they start laying and before predation starts.

## 5.8 Overwintering nucs

Nucs, preferably 14 x 12 with 6 frames, for the overwintering of new queens should be bursting with bees by the time predation starts. Group them together for extra security (see section on defence in the apiary).

## 5.9 Strong colonies going into winter

*"In the life of the honey bee there are only two seasons – winter and preparing for winter"*
*André Blatier, Maître Apiculteur, Bellevigne-en-Layon*

To ensure that colonies survive the winter and have enough adult bees to support the spring surge in brood population, a strong healthy population of winter bees needs:

- A productive young queen
- Well-stocked fat bodies
- Low varroa load
- Access to plenty of stores

Giving colonies syrup and pollen throughout the period of predation, and into autumn, stimulates queens to continue laying in late summer and early autumn and produce good winter bees. See 6 below.

# 6
# F2F - Well-Fed Stock

*"An army marches on its stomach"*

*Frederick the Great, King of Prussia 1740-1786*

*Drawing by Toby Melville-Brown*

As with a successful army, good nutrition is an essential part of any beekeeping operation. It is even more important if the colonies are under threat from Asian Hornets and there are times when they can't fend for themselves.

This section covers different feeding options throughout the year.

## 6.1 Sugar syrup and fondant

Sugar in the form of nectar is an essential component of the diet of adult bees and larvae. It provides energy for internal bodily functions, for work inside and outside of the hive, maintaining the temperature of the brood nest all year round, for generating heat when making wax and building comb and more.

**Spring** We have already seen that we need to build up the colony early to have a big foraging force for the spring flow. We feed syrup to simulate a nectar flow which encourages the queen to increase her laying rate and, after comb change, to stimulate the bees' wax glands for comb building.

Syrup can be given if daytime temperatures are regularly above 10°C. A ratio of 1Kg of white granulated sugar to 650 ml of water is a one-size-fits-all solution.

**Summer and Autumn during the period of predation** Syrup is given continuously to replace loss due to reduced foraging, to preserve or top up their existing stores and to generate water by evaporation (see the section on water below), to build up stores and for queens to produce strong winter bees with well-stocked fat bodies.

**Winter** Fondant is given as usual to maintain a reserve supply of food.

Honey bee foraging for nectar

## 6.2 Pollen

Pollen is used by bees for the following:

- To stimulate the queen to lay eggs.
- To develop young adult bees' hypopharyngeal and mandibular glands which produce brood food and royal jelly, and enzymes in older bees when processing nectar into honey.
- To stimulate the development of their wax glands when building comb.
- To provide protein content in brood food and royal jelly.
- To mix with nectar when making bee bread.
- To build up fat bodies which sustain bees in winter and enable older bees to feed spring larvae.
- For growth and repair of tissue

In our *Fit2Fight* management system we feed pollen:

**Early Spring** - to stimulate the queen to start or increase laying for rapid colony buildup.

**During predation in late Summer and Autumn -** to replace loss from reduced foraging and to keep the queen laying.

**Late summer** - for queens to produce strong winter bees with well stocked fat bodies.
It can be in the following forms:

Pollen supplement which is a mixture of ingredients not collected by the bees, fortified with natural pollen.

Pollen substitute consisting of ingredients not collected by bees that replace natural pollen.

The constituent parts of each of these are:

| Ingredient | Pollen substitute % | Pollen supplement % |
|---|---|---|
| Fat-free Soya flour* | 75 | 60 |
| Brewer's yeast | 25 | 20 |
| Natural pollen | 0 | 20 |
| Total | 100 | 100 |

* Dried skimmed milk and dried egg yolk can also be used.

## Method

Mix the ingredients with sugar syrup or honey into a thick paste to form convenient sized patties about ½" thick and spread onto waxed paper.

Lay the patties flat on top of the brood frames with the wax paper uppermost.

For many of us, collecting pollen and making our own pollen patties is too much faff. There are plenty of commercial products available, but care should be taken to read the list of ingredients carefully to ensure that the pollen, or pollen substitute, content is high enough to give any benefit.

Honey bee foraging for pollen

## 6.3 Water

Water is used by bees as follows:

- Maintaining the fluid balance level of adult and brood haemolymph.
- Diluting stored honey to 50% , the point at which the bees can digest it.
- Controlling temperature of the hive, for example cooling the hive by evaporation

Bees obtain water from evaporation of nectar when it's being processed into honey, and from foraging. There are no means of storing water in the hive. If the bees are unable to go out foraging, we must provide a source of water for them.

A simple contact feeder over the hole in the crown board or on top of the brood frames is all that is required.

Honey bee foraging for water

### Hot Weather

Bees use water to cool the inside of the hive by evaporation. They spread drops of water on the surface of the comb and fan it with their wings.

Other bees fan air in and out through the entrance. Carbon dioxide is removed and replaced with cooler fresh air at the same time.

In addition, bees cluster on the outside of the hive, behaviour called bearding, to reduce crowding inside and improve ventilation.

### During predation

The normal temperature regulation system of the hive is disrupted. There is no foraging for nectar or water, it's too dangerous to beard outside in daylight, the hive entrances are reduced, and ventilation is restricted.

We can help the bees to maintain the correct hive temperature, humidity and CO2 levels by

- Providing shade in the hottest part of the day.
- Opening entrances in the evening when hornets have stopped flying to allow cool, fresh night air to circulate and for undertaker bees to remove the dead.
- Feeding water to compensate for lack of water income from foraging or evaporation of nectar.

**Note** *Opening hives for feeding must be carried out in early morning or evening when the hornets are not flying.*

# 7

# Defence

# Preparing the Battleground

*"Ground is the handmaid of victory"*

*Sun Tzu: the Art of War*

Integrated management in the apiary mainly employs measures which are standard good beekeeping practice, but at a higher level of intensity or frequency.

**In this section we'll look at**

- Apiary security
- Reducing temptation
- Physical defences

## 7.1 Apiary security

Discourage other beekeepers from visiting unless necessary, for example if you need help from your bee buddy, mentor or a Bee Inspector. If you do expect guests, check their clothing and equipment are clean and that they apply the same rules of apiary hygiene as you.

## 7.2 Reducing temptation.

Asian Hornets are attracted by olfactory signals (smells). An open hive releases lots of deliciously inviting aromas of wax, honey and bees so it's a clear invitation to prowling hornets looking for food.

Only open the hives for a specific purpose and do quick in and out visits where possible. You don't need to see the queen every time.

Two exceptions are for swarm prevention and health inspections when the frames need to be shaken free of bees and examined very carefully.

Carry out all manipulations that involve opening hives before the predation period starts if possible.

If you do need to open the hives for any reason, it should be done early in the morning or in the evening as Asian Hornets don't fly at night.

## 7.3 Physical Defences

The level of defence will depend on the traffic light early warning system indicating different levels of threat:

**Green:** I'm in an area where no hornet nests or sightings have been reported yet.

- Grow grass round the hives or put a skirt of boards to stop the hornets getting under and behind the hive, and to give the bees somewhere to hide.
- Don't attract attention to the apiary by opening hives or feeding them during the day.
- Group hives together to provide safety in numbers. Colours or shapes painted on the front of the hives help orientation and reduce drifting.

**Amber:** Asian Hornets confirmed in my area.

- Prepare to take off honey crop.
- Get ready to apply varroa treatment.Insert Varroa boards in hives with open mesh floors to reduce the stress on the bees from hornets flying under the floor.
- Place a tangle of twigs and branches in front of hives to keep the hornets at least 1 metre from the entrance, to give flying bees more dispersal space and somewhere to hide. The gaps should be too small for the hornets to fly through but big enough for the bees. This has a similar effect to a muzzle but costs nothing.
- Instead of branches fit muzzles and put out electric harps if you have them. By doing this early the bees will have time to learn their way in and out. For more on defence equipment see 7.4 below.
- Single hives are particularly vulnerable to attack so if you're a one-colony beekeeper you might consider moving your hive to another apiary where it can benefit from being with others. If you do this, pay careful attention to the health and varroa load of your own hive and those of the apiary you're moving to.

**Red:** Asian Hornets have been seen actively hunting in or near my apiary.

- First of all – don't panic. You and your bees are already well-prepared for this, and you are in charge, not the hornets.
- Take photos and report the incursion on the AH Watch App.
- Inform the NBU or your Area/County AH Coordinator.
- Reduce entrances to one bee space.
- Prepare to put out 'decoy' traps, but don't deploy them unless predation is actually happening (see 8 on traps and trapping).
- Prepare to start feeding to compensate for reduced foraging (see 6 on well-fed stock).
- Provide shade and water in hot weather as the colony's normal temperature regulation mechanisms will be disrupted.
- If predation is heavy, think about moving your hives to another site if possible.

- Stand by to assist the NBU or AHAT in tracking the nest if required (see Part 9 on tracking and bait stations below).
- Have fun swatting hornets with your badminton racket!

## 7.4 Defence equipment

HiveGate is designed to make the entrance easier to defend

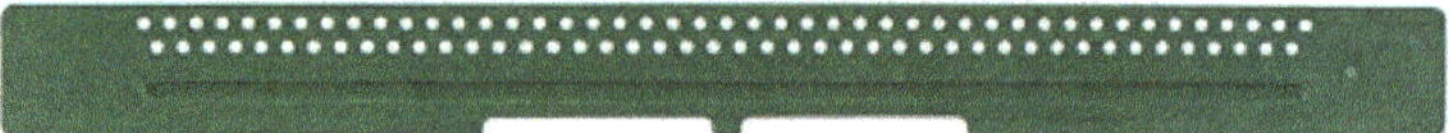

The Nicot entrance reducer is also a mouse guard

**The Muzzle** and the **Electric Harp** have been trialled and used extensively in France and Spain, and have proved successful in reducing stress on the bees.

La Muselière– the Muzzle

*Supplied by E. H. Thorne Ltd.*

"Stop-it"

*Supplied by E. H. Thorne Ltd.*

Muzzles provide a protective screen and a safe place for bees to hide in front of the hive.

La harpe éléctrique

*Supplied by E. H. Thorne Ltd.*

Powered by solar, battery or mains, the electric harp has been found to reduce the pressure of predation on hives, thereby reducing foraging paralysis by a significant amount.

The spaces between the electrified wires are spaced for the bees to fly safely through, and for Asian Hornets to be electrocuted as they touch the wires with their wings.

There is a collecting tray underneath to catch the victims.

For more information on these and other devices, watch the presentation by Asian Hornet expert Andrew Durham at https://shorturl.at/dnqG9

# 8
# Counter Attack
# Salting the Battlefield

*When the Roman Army was victorious, they would throw salt on the battlefield so that the land became useless and the conquered, as a people, were utterly destroyed.*

Roman frieze

We can't expect to utterly destroy the Asian Hornet, but we can reduce the impact that it has on our apiaries. This is how we take the fight to them once they arrive in our apiaries.

## 8.1 Trappin' and zappin'

The subject of trapping is surrounded by controversy, mainly concerning its effectiveness and its impact on other species. It can risk pulling beekeepers in different directions as new thinking continues to emerges and new solutions develop. The suggestions that follow are based on the experiences of those endeavouring to manage Asian Hornets on the Continent and the Channel Islands. Ideas will continue to change as the threat increases in the UK.

Naturally, we all want to be fully pro-active, use the best equipment and techniques available, and protect our bees.

*However, traps are not a silver bullet that will solve the problem of the Asian Hornet in our apiaries – they are just one weapon in our defensive armoury as part of an Integrated Apiary Management Strategy.*

To protect biodiversity, we must avoid catching insects of other species as far as possible. Even if they are released, the experience of being captured and imprisoned with other insects is highly stressful. It can have serious effects from which they may not fully recover. They may not stay alive for very long after release or be able to reproduce.

Whatever the manufacturers claim, as far as we are aware, no trap has yet been produced that is 100% guaranteed to avoid catching innocent victims.

The responsible, measured and proportionate use of traps, with the care necessary to limit by-catch, can be an effective way of slowing down the spread rate of the invasion and reducing the level of predation in our apiaries.

## 8.2 Types of trapping

Baited equipment is currently used in 5 different ways, taking place at different times of the year. The objectives are different, the logistics are different, the execution is different, the outcomes are different, and it's important not to confuse them. They are:

- Spring queen trapping
- Monitoring
- Bait stations
- 'Decoy' trapping in the apiary during predation
- Autumn trapping

Types of trap, bait, record-keeping and disposal of trap contents are also covered in this section.

By-catch in non-selective trap
*Courtesy of le Museum national de l'histoire naturelle*

Wick pots like this one are safe for other species: European Hornet and Speckled Wood butterfly on wick pot

*Courtesy Marc Struye (Belgium) ©*

## 8.3 Spring Queen Trapping

### Background

During the previous Autumn, the majority of Asian Hornets in an original parent nest will have died while large numbers of newly-mated queens (known as gynes) left to find their own spots for hibernation during the winter, usually within about 200 metres from the parent nest.

After the rigours of the winter months, the surviving foundress queens emerge from hibernation anytime between late February to late May. These queens are each looking for a secure, sheltered place to build their first embryo or primary nest near sources of food and water located within an average of 700 metres from the parent nest.

Embryo nest.

The emerging queens are intensely competitive, fighting each other to the death with high attrition rates. This mortal combat between them for nesting sites is known as usurpation.

If she survives this far, each young foundress queen will be weakened and hungry, and must feed herself before she can lay her first eggs. As head of a single parent family, she has sole responsibility for going out to find food for herself and her offspring. She can't travel far, or the eggs and larvae will die of cold. She must feed and return to the nest as fast as possible. Hence, although young queens have been known to forage up to 1km from the primary nest, in practice a radius of 600 metres is more likely.

Queens will continue to forage and be vulnerable until the end of May when the first workers emerge and take over the role of foraging and nest expansion, leaving the queen to lay eggs for the rest of her life.

A very small proportion of new queens will fly, be blown, or be transported further afield, in which case they will escape a local Spring trapping campaign.

***Traps are placed in an area*** where active nests were located the previous late September, October or November, or not found until later in the winter.

Spring trapping can protect colonies whose survival was threatened by Asian Hornets the previous year by targeting places where nests were present and those affected apiaries.

Recently, the French National Plan moved from a form of saturation trapping towards a more localised, less intensive approach.

**Threat level: amber and red**

**Aim**: To prevent emerging queens from setting up new nests.

**Method:**

- Traps can be put out in February when daytime temperatures average 12 ∞C for 4 or 5 consecutive days, and removed at the end of May.
- An evenly-spaced network of traps is placed around last year's nest or the apiary.
- There are normally up to 10 traps distributed within a radius of 500 m.
- The ideal distance between the traps is no more than 350 metres.
- The grid structure provides coverage but may be adjusted to target sources of forage and water.
- Traps are set up in the sun facing south or south-east on the upwind side, in an area around the old nest or the apiary as shown below:

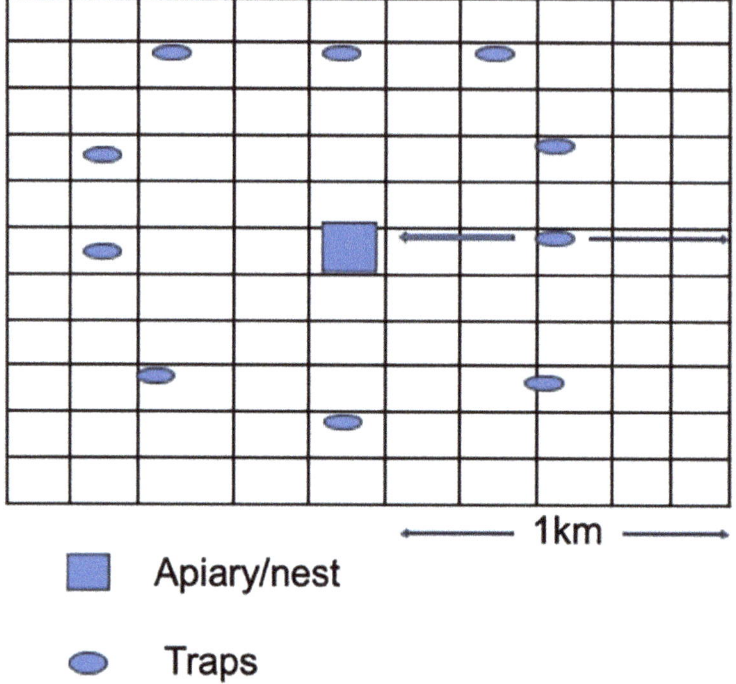


- Traps must be visited frequently to release innocent bystanders and refresh the bait.

**Pros and cons.** May be effective with sufficient access, time and resources to cover a large area.

## 8.4 Monitoring

**Threat Level:** All

Aim: To identify the presence of Asian Hornets (including emerging queens plus migrating, hitch-hiking or wind-blown individuals of all castes).

**Method.** Stations should be placed in locations away from the apiary where surveillance is easy, visually or by camera, and bait can be refreshed. For example, a monitoring station outside the kitchen or office window is an ideal place.

Wick or dish bait stations should be used(see 8.4 below). If traps are deployed instead, they must be of a type that minimises by-catch.

Monitoring can be carried out from late February until late Autumn.

If Asian Hornets are seen, zap them with the Asian Hornet Watch App to alert the NBU, and contact your local Area/County Asian Hornet Coordinator. Enhanced surveillance in the area or track and trace operations may then be put in place.

**Pros and cons.** Potentially a cost-effective early warning system without by-catch.

## 8.5 Bait stations

**Threat level:** Red.

**Aim:** To attract Asian Hornets for monitoring, and track and trace operations.

Method: The bait station is open and available for the Asian Hornets to visit freely. During track and trace operations they are then marked, either while feeding or briefly captured, before returning to the nest.

Usually very simple, bait stations are made out of everyday household objects such as jam jars or take-away boxes and bait-soaked sponges or J-cloths.

**Pros and Cons:** Cost effective, multi-purpose and avoids by-catch.

Asian Hornets on bait station

*Courtesy Marc Struye (Belgium) ©*

## 8.6 'Decoy' trapping during predation in the apiary

**Threat level.** Red.

**Aim.** To divert hornets and so reduce the level and stress of predation.

**Method.** Selective traps are placed near the hives to divert hawking hornets from the foraging bees. Decoy traps should only be set when predation has actually started, otherwise they increase risk by attracting hornets to the apiary.

The traps should be placed at least 1.5 metres from the nearest hive to prevent the Asian Hornet pheromones from stressing the bees. A typical layout would be:

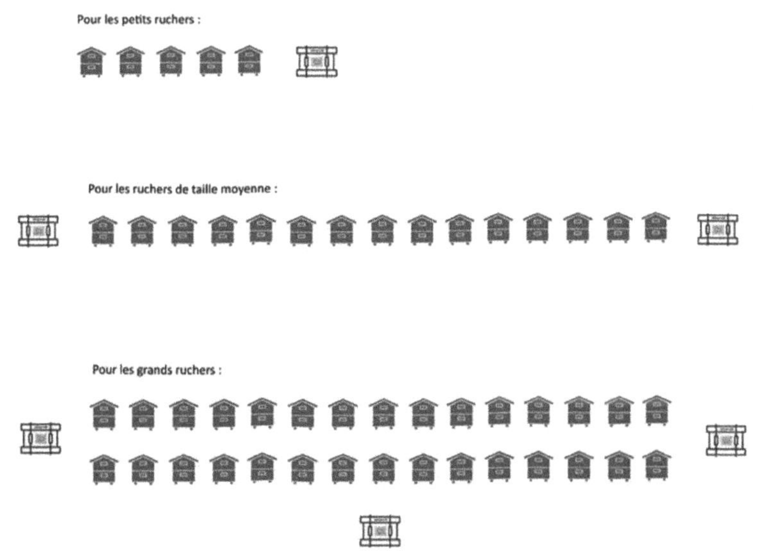


**Pros and Cons.** Decoy trapping can reduce stress on the bees, but is unlikely to have any impact on the number of nests the following year.

## 8.7 Autumn Trapping

**Threat level:** Amber and red

**Aim:** To catch newly-mated queens (or gynes) before they hibernate and so reduce the number of foundress queens that establish new colonies the following Spring.

**Method:** An Autumn trapping campaign requires the same resources and effort as in Spring.

**Pros and Cons**

Autumn trapping is the most favoured by those wishing to minimise by-catch because there are fewer other species on the wing.

However, there seems to be no evidence that Autumn trapping makes any difference due to their high natural attrition rate in Winter and Spring.

Asian Hornet feeding on autumn-flowering ivy

*Courtesy Marc Struye (Belgium)*

## 8.8 Types of Trap

Trap design is still developing and there is much debate about what works best.

Views may change as the UK situation progresses, but naturally we all want to be proactive and source the best equipment and techniques available to protect our bees.

Current thinking is that traps should have a **7-8 mm diameter entry** which is large enough to admit and retain the Asian hornet, but small enough to exclude bigger insects (e.g. European hornets and butterflies). The **exits should be of 6.5 mm**, at the highest point of the trap, to let as many smaller, non-target species escape as possible.

Popular traps are the 'grill and cone type 'such as the *Jabeprode* or Gard'Apis, whereas bottle-shaped designs may be less effective in reducing by-catch. The Gard'Apis trap has two different sized entrance nozzles: the 8 mm red nozzle is designed to catch queens in the Spring and Autumn; and the 7 mm orange nozzle for workers in Summer.

Other, less expensive, traps such as the VetoPhar are useful when adapted to ensure there are 6.5 mm escape holes at their highest point.

More versions are available commercially, and some are being produced by individual or groups of beekeepers who have access to 3-D printing.

## 8.9 The Great Bait Debate

The ideal bait is one that attracts Asian Hornets but is repulsive to all other species. Despite extensive research in Europe and Asia, no such bait has yet been identified, although trials with pheromones are looking promising.

In France a mixture of equal parts of pressed fruit juice, beer and dry white wine is widely used. For the cost-conscious, fermented honey or wax cappings seem also to be enjoyed by the hornets. On Jersey the commercial wasp attractants produced by *Trappit* have been effective in all seasons. Liquid bait should be soaked in a sponge or pot with a wick to prevent insects from drowning.

But by far the best attractant is the smell of other hornets. Once there are hornets in the traps, even after they are dead, they are a powerful attractant to others.

## 8.10 Disposal of Trap Contents

Traps should be emptied after dusk when Asian Hornets have stopped flying. Put the trap in a freezer for 30 minutes to stun the contents. Using a spoon, transfer all but 3 Asian Hornets to a plastic bag and return it to the freezer for 48 hours to kill them. The remaining 3 Hornets can be returned as bait to the unrinsed trap ready for redeployment.

## 8.11 Record Keeping

In addition to the normal hive records, it's essential to keep notes about traps and bait stations: locations; date/times of checks; trap type; bait type; catch; actions taken.

Mobile apps such as Epicollect, Google My Maps and the Hampshire AH Tracker are ideal methods for recording and sharing information with the NBU, local AHATs and other beekeepers about the location of monitoring stations, sightings, nests etc.

# 9
# Finding and Destroying Nests

*"To reduce the amount of predation you have to find and destroy the nests."*
*Alistair Christie Asian Hornet Coordinator States of Jersey*

Although nest destruction alone is unlikely to solve the Asian Hornet problem once a population is established, it can help to slow down the speed of the invasion, reduce predation on apiaries and relieve the pressure on biodiversity in the vicinity.

You can never be sure to have found all the nests, and other new insects may be arriving all the time either migrating, blown on the wind or as hitch-hikers.

Asian hornet nest removal is a dangerous operation and is a job for properly trained and equipped professionals. On no account should you approach a nest or try to remove it yourself.

Whilst the hornets are not aggressive when out hunting, they become very dangerous to anyone approaching too closely to or disturbing the nest. They can attack mob-handed and each inflict multiple, deep and potentially fatal stings.

## 9.1 Track and trace

The most effective way of locating nests is to **catch, mark and release** hornets using bait stations (see 8 above). A bait station is located in an area where Asian Hornets are believed to be flying. The hornet is attracted to the bait and, while feeding may be marked to allow for later identification. When the hornet flies off, its direction is plotted using a compass. Its departure time is also recorded so that, when the identifiable hornet, returns to the bait station, the total intervening time can be calculated. Recognizing that Asian Hornets fly at a rate of 100 metres per minute and are likely to spend very little time in the nest before returning, the approximate distance to the nest can be calculated. Combining the distance calculation plus the direction gives a line to the nest. This process is repeated at two other bait stations some distance away to triangulate and pinpoint the position of the nest.

The hornets are marked with a queen marker pen or may have a tinsel ribbon glued to the thorax. Experiments using miniature radio transmitters attached to the hornets may bring more speed and accuracy in the future.

Asian Hornet on a wick bait station.

Asian hornets on a bait station

Radio tagging in Belgium

*Courtesy of Marc Struye (Belgium)*

At the time of writing, invasive species can only be caught and released by licensed operators, in the case of the Asian Hornet this is limited to members of the National Bee Unit, assisted by local AHAT volunteers. It is hoped that suitably trained volunteers may be authorized to operate independently of the Bee Inspectors in the future.

Destruction of a nest using a lance

*Courtesy Marc Struye (Belgium) ©*

# 10
# Conclusion

After a long 'phony war', the much- feared arrival of the Asian Hornet on our shores has finally come to pass.  Suddenly beekeepers are faced with new challenges that may at first seem overwhelming.

However, all is not lost. By accepting that we have to live with the Asian Hornet, deciding we will not be beaten by it, and by understanding more about it, we can learn how to manage the problem.

This may mean that we have to sharpen up our beekeeping skills and adapt our beekeeping methods to mirror the life cycle of the hornet.

With a phased, targeted response to the threat and by adopting the 3 pillars of Fit2Fight

- **Healthy bees**
- **Strong colonies**
- **Well-fed stocks**

We can manage the situation, reduce the level of predation and relieve the stress of predation on our apiaries.

With a positive attitude, and using the weapons at our disposal, we can continue to enjoy our beekeeping, achieve good outcomes for our bees, and look forward to sharing the wonderful produce of the hive.

# Annex

# FIT2FIGHT CALENDAR

Approximate timing of activity in the apiary depending on location and weather:

# FIT2FIGHT CALENDAR

**Approximate timing of activity depending on location and weather**

| MONTH | AH ACTIVITY | ALERT STATE | ACTION IN APIARY |
|---|---|---|---|
| **January** | Queens in hibernation | Green | Check stores and add fondant if necessary |
| **February** | Depending on weather queens emerge from hibernation & build embryo nest | Green | Feed pollen & syrup to stimulate queen to lay<br>Move hives closer together, mark entrances to reduce drifting.<br>Put out monitoring stations<br>Report sightings on AH Watch App |
| **March** | Queens continue to emerge.<br>Building primary nests<br>First workers emerge | Green | First spring inspection<br>Remove mouse guards, change floors<br>Health inspection<br>Check varroa loads and treat if >1,000 per hive<br>Maintain monitoring stations<br>Feed pollen & syrup to encourage queen laying and worker wax gland development for comb building.<br>Unite weak colonies.<br>Inspect weekly for growth and swarm prevention.<br>Consider shook swarm depending on weather. |
| **April** | Queens continue to emerge | Amber | Health inspection.<br>Maintain monitoring stations<br>Start queen rearing preparations<br>Carry out splits to make increase if required<br>Equalize colonies for simplicity of management<br>Inspect weekly for growth, take samples for Nosema testing if growth is slow.<br>Swarm prevention<br>Ensure queens have enough room to lay<br>Move full frames to outside and empty frames near centre<br>Add supers |
| **May** | Queens continue to emerge. | Amber | Queen rearing begins<br>Inspect weekly for growth and swarm prevention.<br>Add extra brood box to give more space for queen to lay and bonus of getting drawn comb<br>Add more supers<br>Maintain monitoring stations<br>Let grass grow round hives or fit boards round sides |
| **June-July** | Rapid growth of colony.<br>Building secondary nest | Amber | Queen rearing ends.<br>Maintain monitoring stations<br>Inspect weekly for growth and swarm prevention.<br>Move mating nucs to full brood box or 14x12 nucs.<br>Unite weak colonies. Feed if no flow<br>Drone brood uncapping for varroa control<br>Remove honey crop, leave min 1 full super and a half to give bees room<br>Treat for varroa. |
| **Mid July** | Predation period starts<br>Colony building to maximum strength | Red | Final manipulations completed.<br>Reduce entrances to 1 beespace only if predation taking place<br>Feed syrup, pollen, water early morning or evening<br>Provide shade if needed<br>Put woven tangles of branches in front of hives or fit muzzles and deploy harps if you have them<br>'Decoy' Traps only during predation<br>Open entrances at night |
| **August** | Predation period | Red | Feed syrup, pollen, water early morning or evening<br>Keep entrances closed to 1 beespace only when predation in apiary<br>Provide shade if needed<br>Decoy Traps only when predation in apiary<br>Open entrances overnight |
| **September** | Predation period | Red | Feed syrup, pollen, water early morning or evening<br>Keep entrances closed entrances to 1 beespace only when predation in apiary<br>Provide shade if needed<br>Decoy Traps only when predation in apiary<br>Open entrances at night |
| **October** | Predation period<br>Sexuals produced<br>Mating takes place | Red | Feed syrup, pollen, water early morning or evening<br>Open hive entrances if all clear.<br>Open entrances at night<br>Fit mouseguards<br>Consider green woodpecker prevention<br>Fit insulation if not already done |
| **November** | Colony and old queen die<br>Nest abandoned.<br>New queens to hibernation | Amber | Open hive entrances if all clear.<br>Heft and feed foundation if necessary |
| **Deecember** | Queens in hibernation | Green | Treat with Oxalic acid. Heft and feed foundation if necessary |

# Acknowledgements

With grateful thanks to Sarah Bunker and Helen Tworkowski for their combined wisdom and for pointing out my numerous errors and shortcomings.

And especially to my most exacting critic, my wife Penny Melville-Brown OBE.

Any mistakes or omissions in this book are entirely my own.

Writing a book about bees can be a minefield. There's an old beekeeping saying:

*"If you ask 5 beekeepers a question you get 6 different answers."*

How many answers do you get if you ask 5 beekeepers and a lawyer?

# Source Material/References


Basterfield D. Davis I. (2019) British Beekeepers Association

Baxter A. https://www.alanbaxtersblogs.co.uk

Bunker S. 'The Yellow-Legged Asian Hornet - Biology, Spread and Control'. (Published 2024).

Durham A. https://shorturl.at/dnqG9

GDS France

Inventaire National du Patrimoine Naturel INPN https://shorturl.at/htXY8

Mitchel D. University of Leeds https://shorturl.at/dmwAS

National Bee Unit https://www.nationalbeeunit.com/

Stainton, K. (2022) *Varroa Management: A Practical Guide on How to Manage Varroa Mites in Honey Bee Colonies*, Northern Bee Books.

www.ingramcontent.com/pod-product-compliance
Lightning Source LLC
LaVergne TN
LVHW060632110826
845147LV00014B/899

* 9 7 8 1 9 1 4 9 3 4 7 4 2 *